Modern Pulse Diagnosis
Mobile ECG Based

Galina V. Roofener

L.Ac., L.C.H., A.P., Dipl. Ac (NCCAOM)®, Dipl. C.H. (NCCAOM)®

Copyright ©2019 by Galina V. Roofener. All rights reserved.

No part of this publication may be reproduced, stored in a retrieval system, or transmitted in any form or by any means, electronic, mechanical, photocopying, recording or otherwise, without the prior written permission of the author. Permission may be obtained directly from Galina V. Roofener by email GR@AsianTherapies.org.

Title: Modern Pulse Diagnosis Mobile ECG Based
Author: Galina V. Roofener
Edited by: Shellie L. Rosen

ISBN 978-0-359-72890-9

Acknowledgement

To everyone who has inspired my dreams and supported my spirit throughout my journey, I thank you.
To my mom, Vera Veresciak, I extend my highest respect and enormous gratitude for teaching me ways of thinking and taking responsibility for myself and my actions.
To my 'better-half' Brian Terry, I am grateful for your encouragement and endless patience as I created this book.
To Shellie Rosen, your encouragement and assistance converting and editing my "Ruglish" to English has been a wonderful gift.
To all my patients and colleagues, thank you for providing me with the continuous learning opportunities that made this book possible.

Galina V. Roofener

Table of content

1. Traditional East Asian Medicine pulse diagnosis overview — 7
2. Modern technology in pulse diagnosis — 13
3. Basic principles of an Electrocardiogram (ECG) — 15
4. Red flags that require a referral to a Medical Doctor (MD) — 18
5. ICD 11 development and the standardization of terminology — 24
6. Clinical TCM Pulse Differentiation — 30
7. East – West Mobile ECG Based Pulse Interpretation — 33
8. Mobile ECG Based Pulse Case Studies — 37
9. References — 61

Chapter 1

Traditional Far East Asian Medicine Pulse Diagnosis Overview

Pulse diagnosis is one of the four pillars (examinations) in East Asian medicine diagnosis. The importance of pulse palpation and its meanings has been discussed in many classical texts of Traditional Chinese Medicine (TCM) starting with *Huang Di Nei Jing* written circa 200 B.C.

The importance of pulse diagnosis has waxed and waned since its emergence in Chinese medicine. Perhaps the most influential text thus far has been the *Bin Hu Ma Xie*, written in 1518 by a prominent Chinese scholar named Li Shi Zhen. This text remains an essential reference in the modern study of TCM.

The primary importance of pulse diagnosis is in its ability to reflect the state of Qi and Blood throughout the body. The main drawback to the widespread use of pulse diagnosis is in the subjective nature of the traditional reading method. Reading a pulse traditionally relies upon extensive practice guided by a skilled teacher, which can be challenging to attain and master. Additionally, a pulse is quickly changing variable, subject to many external and internal influences.

Factors that can influence pulse include:
- Activity level (sleeping, resting, exercise)
- Gender
- Fitness level
- Air temperature
- Body position (standing up or lying down)
- Emotions
- Body size
- Medications
- Food and drink
- Illness

The diverse population within the United States presents a complex framework for practitioners learning the craft of pulse diagnosis. Determining pulse depth depends on the size and body build of a patient. The amount of soft tissue on the top of *chi* (proximal) position can vary by half of an inch or more. This tissue variance can create a false impression of pulse depth.

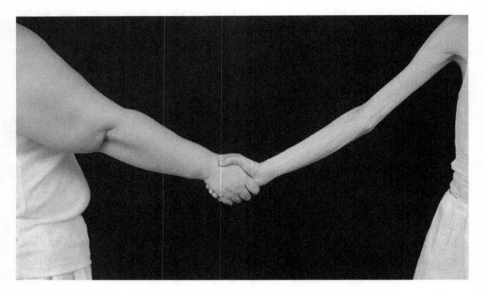

In ancient times TCM master Li Shi Zhen described calculating a pulse rate by counting the number of beats per one of the physician's breaths. An average adult is expected to have a respiration rate of 12-18 breaths per minute (at rest). The beats amount to 48 and 72 per minute.

Contemporary TCM practitioners, now use time in the place of respiration when counting the pulse. A pulse rate of 60-80 beats per minute is within a healthy range. It seems as though the current pulse rate is faster than those reported from the time of Li Shi Zhen.

The current Western medical standard for a regular pulse is between 60 and 100 beats. Tachycardia is a heart rate above 100 beats per minute, and bradycardia is a pulse rate below 60 beats per minute. An athletic individual may present considerably lower heart rates. For example, a runner's resting heart rate may register below 50 beats per minute and still be considered normal.

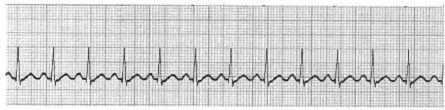

Tachycardia.

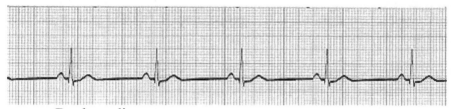

Bradycardia.

In TCM a rapid heart rate is often considered a sign of heat, and a slow heart rate is considered a sign of cold. Comparison of the ancient Chinese heart rate range of 48-72 per minute to the modern American heart rate range of 60-80 per minute poses the question: Were ancient people more athletic, or are modern humans overheated?

Perhaps the answer is 'true' for both. Modern life has created a shift allowing many people to avoid hard physical labor and more sedentary professions. Many patients exhibit signs of deficient heat that manifest as a skyrocketing number of chronic inflammatory, metabolic, and autoimmune diseases among many others.

The palpation of the pulse is an essential diagnostic method to determine the state of the Qi. East Asian medicine describes "Qi" as a moving force that separates dead from live matter. A patient that has no Qi does not exhibit a pulse. The concept of Qi has been difficult for the Western medical community to grasp. The majority of acceptable diagnostic methods in Western medicine are based on the visible material structure. Diagnostics may include histology, X-Ray, or a chemistry-based laboratory analysis. Qi is electromagnetic and shall be investigated and measured with the tools of physics.

Dr. Daniel Keown, author of "The Uncharted Body" writes the most scientifically, embriologically, anatomically and physiologically correct explanation of Qi. Keown notes Qi, as a piezoelectric current, which flows through electrically conductive collagen fibers. These fibers are immersed in lymphatic fluid and organized into myofascial meridians, which penetrate the entire body.

A pacemaker generates electrical signals that command the heart chambers to contract. This signal propagates through the arterial wall collagen lining commanding muscles of the arterial wall to contract. This contraction continues the propulsion of blood throughout the extensive arterial circulatory system.

It is because of these mechanisms that a human heart, the size of a fist can move blood great distances throughout the body. TCM explains the phenomenon of blood propulsion through the body stating, "Qi is the commander of Blood."

Simultaneously, when the body is blood deficient, the tissue is dry. Electrical conductivity is minimal in dry collagen fibers, resulting in a weakened electrical impulse. In physics this concept is illustrated wherein electrical conductivity is enhanced with the introduction of water. The TCM postulate, "Blood is the mother of Qi" also reflects the relationship between moisture and conductivity.

The signal the practitioner receives at the radial artery during pulse palpation refers to: the volume of blood and the strength of propagating electrical contractions of the arterial wall. The sensation manifests similarly to circular waves created from a stone thrown into water. The pulse has three prominent waves (P, R & T) passing through each position. Their presence and strength explain the difference felt at each of three positions.

Many modern technological tools for detecting and measuring Qi through meridians are available on the market. One of the oldest and well-researched is, a Ryodoraku system, based upon the measurement of low electric resistance potential at yuan points on meridians. This technique was brought forth in the 1950s; however, it has not been a widely adopted method in the west. An ECG is considered a prominent scientific test in the west to detect the electrical activity of the heart. An ECG is a readily available test that can also be used to measure and interpret the state of Qi in the human body.

This is an electric impulse commanding heart chambers to contract that generates the pulse. The strength of the electric impulse is visible in the amplitude of a wave spike. Here, one can determine the strength of Qi by the height of the R wave spike that is generated. An ECG report can help a practitioner determine heat or cold conditions and other factors as will be discussed in detail in subsequent chapters.

The intricate aspects of pulse diagnosis (based upon the personal experience of practitioners) can be difficult to standardize. Measurable outcomes that are reproducible by practitioner peers can provide higher credibility and increased acceptance from modern medicine. This credibility may outweigh the loss of some intricacy.

A diagnosis should never be made based on a single diagnostic test, be it pulse palpation, blood biochemistry, tongue observation, or MRI. Any test MUST always be correlated with the clinical presentation and confirmed by multiple examinations. If a single test does not correlate with other findings, the correctness of the test is doubtful, and the test should be repeated.

This book is intended for individuals familiar with the theory and clinical practice of East Asian medicine. This educational material is based on the clinical expertise of the author. This material aims to offer a modern method of recognizing and registering TCM pulse readings to add to the clinical findings. Practitioners must draw conclusions based on their knowledge, experience, clinical observations, and scope of practice as outlined by law. Patient referrals are recommended whenever patient symptoms or pathology extend beyond the individual practitioners reach.

The information in this book is not intended to diagnose or treat any western medical conditions. ECG interpretation described in this book is not advised as a substitute for western medical care of any disease, specifically cardiovascular concerns.

Neither the author, editors nor contributors to this text assume any liability for injury and damage to persons or property from the use or operation of any methods, products, instructions or concepts referred to or contained within this publication.

Chapter 2

Modern Technology in Pulse Diagnosis

Telemedicine is a modern health technology with great promise for traditional medicine practitioners. Traditional medicine practitioners have the potential to increase their patient load, serving a much larger patient population. Patients who live in remote locations benefit from access to traditional Chinese medicine (TCM) treatments.

Traditional Chinese herbal medicine (TCHM) consultations are the most practical way to offer benefits to patients in a virtual setting. Through secure HIPAA compliant virtual portals patients and practitioners exchange information that allows for an herbal consult. The caveat? How does a practitioner gather objective data such as tongue observation and pulse palpation?

The existing research that focuses on the development of electronic devices, is designed to reproduce the pressure of a human fingers measuring a pulse. These measurements are intended to read the complexity of a TCM pulse according to theory with three depths, six locations along with other qualities descriptions. Even if these devices are useful in their interpretation outcomes, the cost alone makes them prohibitive for extensive use.

Recent developments in mobile electrocardiogram (EKG or ECG) devices and accompanying apps have made them available for purchase by the general public. This provides a convenient and affordable device for practitioners to perform an objective clinical measure. An ECG can be taken by the patient at home and transmitted to a practitioner. This book intends to describe TCM pulse interpretation based in mobile ECG graph.

ECG taking directions for patient to follow:

- ECG readings should take place in a calm, restful environment.
- Allow for 1 hour of rest after strenuous physical activity, or after eating large meal.
- Allow 15 minutes after urination, defecation or ingestion of liquids.
- Do not take an ECG immediately after a stressful emotional event; allow yourself to calm down completely.
- Do not talk during an ECG reading.
- Hold your fingers steadily upon electrodes, then relax, and proceed.

WARNING: ECG tracing is affected by patient motion. Shivering and tremors, anxiety, or patients in uncomfortable positions can create the illusion of cardiac arrhythmia. An ECG may contain artifacts that are distorted signals caused by secondary movements internal or external (such as muscle) sources cause interference with a wearable electrical device.

Chapter 3

Basic Principles of ECG

An electrocardiogram — abbreviated as EKG or ECG — is a test that measures the electrical activity of the heart. The electrical conduction system of the heart generates an electric impulse. This impulse begins in the right atria and ends in the left ventricle, resulting in blood output in the large blood circulation circle. The heartbeat can be felt as a pulse in a few locations on the body, including the radial artery.

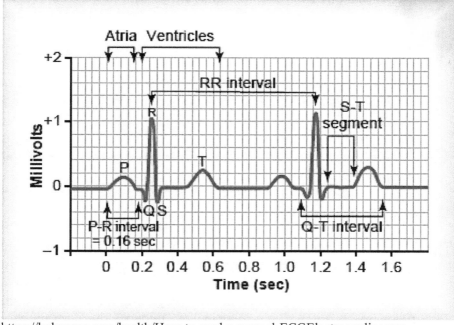

https://hubpages.com/health/How-to-read-a-normal-ECGElectrocardiogram

Graph paper print outs illustrate the measurements of the amplitudes and intervals of an ECG at a standard scale: 1 mm (one small box on the standard ECG paper) represents 40 milliseconds on the x-axis and 0.1 millivolts on the y-axis.

The P wave and PR intervals represent electrical impulse flows through the atria. P wave shape changes:

15

- Peaked, notched, enlarged or inverted may indicate pulmonary disease, pulmonary embolism, enlargement of atria, or valve issues.
- Reversed, absent or located too far from a QR peak may indicate the sinus node is blocked, ischemic infarction, or severe hyperkalemia.

Typically, a large *right atrium* shows up as a tall, peaked P wave while a large *left atrium* looks like a two-humped bifid P wave.

<u>The QRS complex</u> represents an electric impulse flow through the ventricles. The electric impulse causes ventricles to contract and push blood into the arteries. This discharge is what creates the pulse.

The Q wave is a downward spike, the R wave is an upward spike, and the S wave is a second downward spike. The QRS wave shape changes:
- A weak wave may mean premature ventricular contraction (PVC).
- Absent may mean no ventricle contraction.
- A wide QRS complex may mean a right or left bundle branch block, ventricular flutter or fibrillation, hyperkalemia or tricyclic antidepressants overdose.
- An unusually tall QRS complex may represent a left ventricular hypertrophy.
- A very low-amplitude QRS complex may represent a pericardial effusion or an infiltrative myocardial disease.

<u>The ST segment</u> represents a relaxation of ventricles. The ST segment shape changes:
- An inverted ST segment may mean myocardial ischemia, digoxin toxicity, hypokalemia, hypomagnesemia, acute posterior myocardial infarction, ventricular hypertrophy, pulmonary embolism, hyperventilation, bundle branch block or normal variant.
- A peaked ST segment may mean myocardial ischemia, acute myocardial infarction, left bundle branch block, possibly acute pericarditis, hyperkalemia, pulmonary embolism, hypothermia, digoxin toxicity, left ventricular hypertrophy or a normal variation (e.g., athlete's heart).

<u>The T wave</u> represents ventricular repolarization. T wave shape changes:
- A tall and peaked T wave may mean hyperkalemia, hypocalcemia, and left ventricular hypertrophy, left bundle branch block, stroke.
- A depressed T wave may mean myocardial ischemia, myocarditis, hyperventilation, anxiety, drinking hot or cold beverages, left ventricular hypertrophy, certain drugs toxicity, pericarditis, pulmonary embolism, conduction disturbances (e.g., right bundle branch block), electrolyte disturbances (e.g., hypokalemia) or age and race variances.

<u>U wave</u> represents repolarization of His–Purkinje fibers, it may not appear on an ECG; if it does appear it is upright and rounded wave that follows the T wave. U wave shape changes:
- If it is very prominent may mean hypercalcemia, hypokalemia, hypomagnesemia, hyperthyroidism, ischemia, or digoxin toxicity.

<u>QT interval</u> represents the time needed for ventricular depolarization and repolarization. QT interval changes in shape:
- A long QT interval may indicate myocardial infarction, myocarditis, hypocalcemia, hypokalemia, hypomagnesemia, hypothyroidism, subarachnoid or intracerebral hemorrhage, stroke, congenital long QT syndrome. A long QT interval may also be due to antiarrhythmics (e.g., solatol, amiodarone, quinidine), tricyclic antidepressants, phenothiazines, and other drugs.
- A short QT interval may indicate hypercalcemia, hypermagnesemia, Graves' disease, or digoxin toxicity.

RED FLAG: Prolongation of the QT interval may lead to a life-threatening ventricular arrhythmia or fibrillation, also called torsade's de pointes, can result in sudden cardiac death.

Chapter 4

Red Flags That Require a Referral to a Medical Doctor

Ventricular fibrillation: according to the Merck Manual: *"Torsade's de pointes* is a specific form of polymorphic ventricular tachycardia in patients with a long QT interval. It is characterized by rapid, irregular QRS complexes, which appear to be twisting around the ECG baseline. This arrhythmia may cease spontaneously or degenerate into ventricular fibrillation. It causes significant hemodynamic compromise and often death. Diagnosis is by ECG. Treatment is with IV magnesium, measures to shorten the QT interval, and direct-current defibrillation when ventricular fibrillation is precipitated."

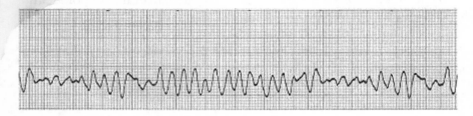

Early signs and symptoms of ventricular fibrillation:
- Chest pain
- Tachycardia
- Dizziness
- Nausea
- Shortness of breath
- Loss of consciousness

Risk factors:
- Age
- Gender
- A previous episode of ventricular fibrillation
- A previous heart attack
- Congenital heart disease/ Cardiomyopathy
- Hepatic dysfunction
- Renal dysfunction

- Significant electrolyte abnormalities
- Starvation
- Liquid protein diets
- Anorexia nervosa
- Hypothyroidism
- Hypothermia
- HIV infection
- Autoimmune connective tissue disease
- Toxic exposure to organophosphate insecticides
- Certain drugs

Example of drugs that increase the QT interval:

Antiarrhythmics
Sotalol
Quinidine
Procainamide
Disopyramide
Flecainide

Antipsychotics
Chlorpromazine
Haloperidol
Droperidol
Amisulpride
Thioridazine
Pimozide
Clozapine

Antidepressants
Amitriptyline
Nortriptyline
Desipramine
Venlafaxine
Bupropion
Moclobemide

Antihistamines
Diphenhydramine
Astemizole
Loratidine
Terfanadine

Antimicrobials
Erythromycin
Clarithromycin
Moxifloxacin
Fluconazole
Ketoconazole

Antiemetics
Domperidone
Droperidol

Others
Methadone
Sunitinib
Hydroxychloroquine
Quinine
Telaprevir

Hyperkalemia (high potassium) is a life-threatening medical emergency. According to the Mersk manual: "ECG changes are frequently visible when serum potassium is > 5.5 mEq/L. Slowing of conduction is characterized by an increased PR interval and shortening of the QT interval. Tall, symmetric, peaked T waves are visible initially. Potassium > 6.5 mEq/L causes further slowing of conduction with widening of the QRS interval, disappearance of the P wave, and nodal and escape ventricular arrhythmias. Finally, the QRS complex degenerates into a sine wave pattern, and ventricular fibrillation or asystole ensues."

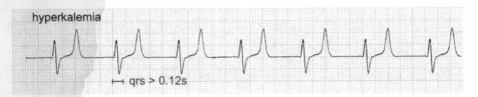

Symptoms (if any) are non-specific and include:
- palpitations
- malaise
- muscle weakness

Causes:
- kidney failure
- Addison's disease or mineralocorticoid deficiency
- hypoaldosteroism
- rhabdomyolysis
- systemic lupus erythematosus
- acidosis
- burns or bleeding into soft tissue or gastrointestinal tract
- insulin deficiency or resistance

Drugs that may increase potassium
- Potassium-sparing diuretics (spironolactone, triamterene, amiloride)
- NSAIDs.
- ACE inhibitors.
- Angiotensin-receptor blockers (ARBs)
- Cyclosporine or tacrolimus.
- Trimethoprim-sulfamethoxazole.
- Heparin

- Antifungals (ketanazole, Fluconazole, Itraconazole)
- Beta-Blockers
- Digoxin
- Penicillin G in high dose
- Fluorosis
- Pentamidine
- Succinylcholine
- Tacrolimus
- Triamterene
- Yasmin (Synthetic Progestin - has Spironolactone like effect)
- Excessive potassium supplementation

Atrial Fibrillation: is the most commonly sustained rhythm disorder observed in clinical practice. It is often not immediately life threatening, but it can have serious clinical consequences such as ischemic stroke. According to the Merck Manual: "Atrial fibrillation (A-fib) is a rapid, irregularly irregular atrial rhythm. Atrial thrombi often form, causing a significant risk of embolic stroke. Diagnosis is by ECG. Treatment involves rate control with drugs, prevention of thromboembolism with anticoagulation, and sometimes conversion to sinus rhythm by drugs or cardioversion. AF is affecting about 2.3 million adults in the US. Men and whites are more likely to have AF than women and blacks. Prevalence increases with age; almost 10% of people over 80-year-old are affected."

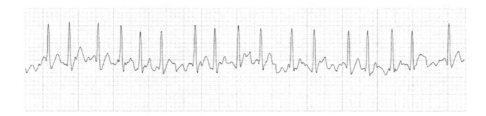

Early signs and symptoms of atrial fibrillation:
- palpitations
- chest pain or vague chest discomfort
- fatigue
- shortness of breath
- weakness

- lightheadedness
- dizziness
- fainting
- confusion
- intolerance to exercise

Risk factors:
- age
- hypertension
- congestive heart failure
- coronary artery disease
- heart valve disease
- hypertrophic cardiomyopathy
- heart surgery
- congenital heart defects
- pericarditis
- hyperthyroidism
- obesity
- diabetes
- kidney disease
- binge drinking
- certain drugs

Drugs that may cause Afib
- antiarrhythmics
- adenosine
- theophylline
- dopamine
- alcohol
- acetylcholine (nicotine, atropine)
- sympathomimetics
- verapamil
- chemotherapeutic agents
- coronary vasoconstrictors
- anti-migraine drugs
- antidepressants
- antipsychotics

- diuretics,
- glucocorticoids
- NSAIDS

Myocardial Infarction: according to the Merck manual: "Acute myocardial infarction is myocardial necrosis resulting from acute obstruction of a coronary artery. A diagnosis is made by performing an ECG and through the presence or absence of serologic markers. Treatment is antiplatelet drugs, anticoagulants, nitrates, beta-blockers, statins, and reperfusion therapy. For ST-segment-elevation myocardial infarction, emergency reperfusion is via fibrinolytic drugs, percutaneous intervention, or, occasionally, coronary artery bypass graft surgery. For non-ST-segment-elevation MI, reperfusion is via percutaneous intervention or coronary artery bypass graft surgery."

Symptoms:
- deep, substernal, visceral ache or pressure, radiating to the back, jaw, left arm, right arm, shoulders, or all of these areas
- dyspnea
- diaphoresis
- nausea and vomiting
- pain may be slightly or temporarily relived by rest or after administering nitroglycerin.
- 20% of acute MIs are asymptomatic or cause vague symptoms
- MIs are more common in patients with diabetes
- women are more likely to present with atypical chest discomfort
- elderly patients may report dyspnea more than chest pain

Chapter 5

ICD 11 Development and Standardization of Terminology

Non-Chinese speaking East Asian Medicine practitioners experience communication difficulties due to variances in translation. The World Health Organization (WHO) undertook the enormous task of standardizing TCM terminology. The introduction of Traditional Medicine diagnosis terminology is within the ICD 11 diagnostic codes, (under chapter 26). Available at: https://icd.who.int

There is a great benefit when East Asian medicine providers can globally agree on a universal language. The sacrifice for this advancement is the requirement of practitioners to learn the new identifying codes. Additionally, not all of the new descriptions are easy to distinguish, which may add to potential confusion. Li Shi Zhen described 29 pulses in his book Pulse Diagnosis, Paradigm Press; 1 edition (February 8, 1993), below are 48 TCM pulse descriptions as outlined in *WHO International Standard Terminologies on Traditional Medicine in the Western Pacific Region*. World Health Organization (2007):

#	WHO code	WHO name	WHO description
1	2.4.25	tranquil pulse	pulse that becomes gentle, in the course of an illness, usually indicating improvement of the condition
2	2.4.26	agitated pulse	pulse that becomes rapid and rushing, usually indicating deterioration of the condition
3	2.4.27	fulminating pulse	sudden throbbing of a hardly perceptible pulse, usually indicating a critical condition
4	2.4.28	pulse bereft of stomach qi	a pulse that has lost its usual rhythm, frequency and evenness, indicating lack of stomach qi

5	2.4.29	floating pulse	a superficially located pulse which can be felt by light touch and grows faint on hard pressure
6	2.4.30	sunken pulse	a deeply located pulse which can only be felt when pressing hard, also called deep pulse
7	2.4.31	slow pulse	a pulse with less than four beats to one cycle of the physician's respiration, the same as bradycardia
8	2.4.32	rapid pulse	a pulse with more than five or six beats to one cycle of the physician's respiration, the same as tachycardia
9	2.4.33	surging pulse	a pulse beating like dashing waves with forceful rising and gradual decline, also called flooding pulse
10	2.4.34	fine pulse	a pulse as thin as a silk thread, straight and soft, feeble yet always perceptible upon hard pressure, also called thin/thready pulse
11	2.4.35	vacuous pulse	a general term for a feeble and void pulse
12	2.4.36	replete pulse	a general term for a pulse felt forceful at all the three sections, cun/inch, guan/bar and chi/cubit, also called forceful pulse
13	2.4.37	long pulse	a pulse with beats of long duration, exceeding cun/ inch, guan/bar and chi/cubit sections
14	2.4.38	short pulse	a pulse with beats of short duration, only felt at guan/ bar section
15	2.4.39	slippery pulse	a pulse coming and going smoothly like beads rolling on a plate
16	2.4.40	rough pulse	a pulse coming and going unsmoothly with small, fine, slow joggling tempo like scraping bamboo with a knife
17	2.4.41	string-like pulse	a straight, long and taut pulse, like a musical string to the touch

18	2.4.42	tight pulse	a pulse feeling like a tightly stretched cord
19	2.4.43	soggy pulse	a thin and floating pulse which can be felt on light pressure, but growing faint upon hard pressure
20	2.4.44	moderate pulse	a pulse with four beats to one cycle of the physician's respiration, even and harmonious in its form
21	2.4.45	relaxed pulse	a pulse with decreased tension
22	2.4.46	faint pulse	a thready and soft pulse, scarcely perceptible
23	2.4.47	weak pulse	a pulse that is deep, soft, thin and forceless
24	2.4.48	dissipated pulse	a pulse that feels diffusing and feeble upon a light touch and faint upon hard pressure
25	2.4.49	hollow pulse	a floating, large, soft, and hollow pulse
26	2.4.50	drumskin pulse	a pulse felt hard and hollow as if touching the surface of a drum
27	2.4.51	firm pulse	a broad, forceful and taut pulse, deeply seated and felt only by hard pressure
28	2.4.52	hidden pulse	a pulse which can only be felt upon pressing to the bone, located deeper than sunken pulse or even totally hidden
29	2.4.53	stirred pulse	a quick, jerky pulse, like a bouncing pea, slippery, rapid and forceful
30	2.4.54	intermittent pulse	a moderate weak pulse, pausing at regular intervals
33	2.4.57	large pulse	a broad pulse with a bigger amplitude than normal
34	2.4.58	soft pulse	a pulse felt softer than normal
35	2.4.59	racing pulse	a pulse having more than seven beats per respiration

36	2.4.60	strange pulse	special pulses signifying critical conditions
37	2.4.61	true visceral pulse	a pulse condition indicating exhaustion of visceral qi
38	2.4.62	pecking sparrow pulse	an urgent rapid pulse of irregular rhythm that stops and starts, like a sparrow pecking for food
39	2.4.63	seething cauldron	an extremely rapid floating pulse that is all outward movement and no inward movement, also known as bubble-rising pulse
40	2.4.64	waving fish pulse	a pulse that seems to be yet seems not to be present, like a fish waving in the water
41	2.4.65	darting shrimp pulse	a pulse that arrives almost imperceptibly and vanishes with a flick, like a darting shrimp
42	2.4.66	leaking roof pulse	a pulse that comes at long and irregular intervals, like water dripping from a leaky roof
43	2.4.67	untwining rope pulse	a pulse, not loose, not tight, with an irregular rhythm like an untwining rope
44	2.4.68	flicking stone pulse	a sunken replete pulse that feels like flicking a stone with a finger
45	2.4.69	upturned knife pulse	a pulse like a knife with the blade pointing upward, i.e., fine, string-like, and extremely tight
46	2.4.70	spinning bean pulse	a pulse that comes and goes away, elusive like a spinning bean
47	2.4.71	confused skipping pulse	a pulse extremely fine and faint, and urgent, skipping and chaotic
48	2.4.72	anomalous pulse	a sudden change of pulse condition in a laboring woman

Pulses most commonly used in United States before ICD 11

Pulse	Description
Fast	Pulse is greater than 80 beats per minute.
Slow	Pulse is less than 60 beats per minute.
Full	Pulse can be felt strongly on all three levels (superficial, middle and deep).
Empty	Feels weak, and with pressure cannot be felt at all.
Floating	Pulse is the strongest at upper level, and can be felt with only a light touch.
Deep	Pulse is the strongest at lowest level and requires deep pressure to be felt.
Knotted	Irregularly irregular and slow. Pulse is missing a beat with no apparent pattern.
Hurried	Irregularly irregular and fast. Pulse is missing a beat with no apparent pattern.
Intermittent	Regularly irregular. Pulse is missing a beat with a definite pattern.
Choppy	Pulse is uneven and rough, feels like a knife scraping bamboo.
Slippery	Pulse feels like pearls rolling on the dish. This pulse quickly hits each position and quickly rolls away.
Thin/ Thready	Pulse feels thinner than it should.
Tight	Pulse feels taught like a rope, thicker than wiry, feels as if the pulse evenly hits the fingers in different places with every beat.
Wiry	a straight, long and taut pulse, like a musical string to the touch

Summary of Pulses according to 8 principles differentiation

Principles	Category	Palpation	Similar pulses
Exterior	superficial	Felt with light pressure	Floating, hollow, drumskin
Hot	replete	More than 5 beats per breath (>80 BPM)	Rapid, stirred, skipping, racing, pecking sparrow, seething cauldron, confused skipping
Excess	full	Large, long	Surging, large, long, slippery, string-like, tight
Interior	deep	Felt with deep pressure	Deep, sunken, firm, hidden, flicking stone
Cold	slow	Less than 3 beats per breath (<60 BPM)	Slow, intermittent, bound, leaking roof, untwining rope, spinning bean
Deficient	vacuous	Small, short	Fine, rough, short, soggy, relaxed, faint, weak, dissipated, soft, waiving fish, darting shrimp

Chapter 6

Clinical TCM Pulse Differentiation

"The three qualities for a normal pulse: A normal pulse is regular, smooth, and harmonious, indicating the presence of stomach qi. It is supple and powerful, indicating the presence of vitality. It can be felt on deep palpation, indicating the presence of root." (Source: *WHO International Standard Terminologies on Traditional Medicine in the Western Pacific Region*, World Health Organization 2007).

The centuries-old wisdom compiled in the standard terminology manual can be challenging to apply in modern times. This text attempts to categorize pulses based upon parameters that most significantly influence TCM pattern diagnosis in the outpatient clinical practice. This method of categorization allows for a logical pulse description in clinical records.

This text draws upon the experience of work performed within a major hospital in which five acupuncturists collaborated in patient cases. These East Asian medical practitioners were diverse in language, background, and education from different countries.

Aspects of practice varied among the practitioners such as pulse palpation techniques, clinical record keeping, writing, and comprehension skills. A problematic feature of this sort of collaborative effort is the ability to sort through the many methods and views to arrive at the same diagnostic conclusion the prior practitioner intended to convey in their clinical findings.

Acupuncture has become a number one non-pharmacologic treatment routinely recommended by physicians for pain management. The practice is relying on insurance reimbursement more than ever before. As exposure within insurance evaluation broadens, so too does the scrutiny of clinical records by insurance auditors.

As East Asian Medicine adapts to evolving change, language standardization is imperative. The current United States healthcare system is transforming from a fee-per-service-based model to an outcome-based model. Consistent and clear, measurable progress of treatment outcomes is a priority as the profession moves forward.

TCM attempts to describe the pulse as a multidimensional phenomenon. For the ease of clinical comprehension, each pulse may be described in five dimensions that are the most relevant for a thorough diagnosis.

Pulse descriptive characteristics most relevant in US clinical practice:

- a) **Rate**
  - Racing - above 100 beats per minute
  - Rapid - above 80 beats per minute
  - Moderate - between 60 and 80 beats per minute
  - Slow - below 60 beats per minute

- b) **Force**
  - Replete
  - Smooth
  - Uneven
  - Vacuous

- c) **Level**
  - Floating
  - Average
  - Rootless
  - Sunken

- d) **Rhythm**
  - Regular
  - Regularly irregular
  - Irregular
  - Skipping

e) **Shape**
- String-like (Wiry)
- Slippery
- Thin (Thready)
- Rough (Choppy)

Chapter 7

East – West Mobile ECG Based Pulse Interpretation

1. Pulse levels as reflected on a single lead mobile ECG

Classically the pulse is evaluated on three levels:
- Superficial level - easily felt beneath the skin (associated with the Yang state of the body).
- Middle level - felt at moderate pressure (the level of soft tissue associated with the Blood state of the body).
- Deep level - felt close to the bone (associated with Yin state of the body).

Each ECG starts with a Π, which represents the range of normal. The upper horizontal line represents the upper level of a normal pulse. The bottom horizontal lines represent the deep level of a normal pulse. Therefore, we can determine that R wave spike extending above upper level of Π represents floating level and Q&S spikes extending below Π represent root level.

2. Pulse positions as reflected on single lead mobile ECG

According to Li Shi Zhen, the pulse is evaluated bilaterally on three positions and corresponds to regions of the body:
- *Cun* - is distal position and represents the upper region of the body (Upper Burner) that extends from the diaphragm to the vertex of the head and houses lungs and heart.
- *Guan* - is middle position and represents the middle region of the body (Middle Burner) that extends from navel to diaphragm and houses liver and spleen/stomach.
- *Chi* - is proximal position and represents the lower region of the body (Lower Burner) that extends from navel to feet and houses kidneys.

Looking at an ECG, three distinct positions are observable:

- P wave is generated by the ventricles, which control the pulmonary circuit of circulation. Changes in shape may be reflective of pulmonary disorders. Therefore, T wave is in perfect correlation with the Upper Burner or *Sun* position.

- QRS wave changes in shape may be associated with the use of tricyclic antidepressants. In TCM, depression has a clear pathogenesis of Liver Qi stagnation. TCM describes the physiology of Liver Qi directional movements as strong Qi moving in all directions. Such presentation correlates with a QRS wave that represents ventricular contractions that eject blood in a general circulation delivering blood to the entire body (head to toes). One can assume that a QRS wave represents the *Guan* position.

- T wave shape changes are notably changed in kidney failure, mineral, and corticosteroid deficiencies and can be affected by drugs such as Yasmin (a synthetic progestin). One might assume that a T wave represents a Lower Burner or *Chi* position.

Below is representation of the aspects of the ECG mentioned above.

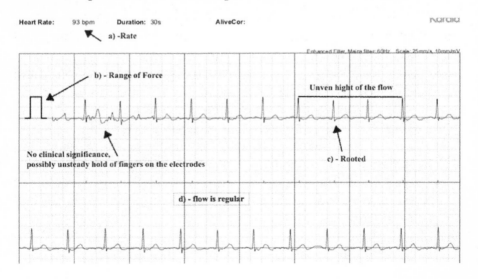

3. Pulse qualities as reflected on single lead mobile ECG

a) **Rate** - is the number of beats per minute. The position on an ECG is along the longitudinal dimension. Rate is clearly displayed on the top of an ECG printout.

Clinical significance:
- Racing – Fire or extreme replete or vacuous heat
- Rapid - replete or vacuous heat
- Moderate – normal or requires clinical correlation
- Slow – cold, or Heart Qi vacuity, or normal (athletes)

b) **Force** - is represented as the height of ECG waves, mostly in the QRS complex. Each ECG starts with the symbol Π, to represent the range of normal. A pulse, which rises above a range of normal, may be classified as replete. R wave rising below one third of the range of normal maybe be classified as vacuous.

Clinical significance:
- Replete - excess, accumulation of pathogen
- Smooth – normal or requires clinical correlation
- Uneven – Qi stagnation, interior Wind
- Vacuous – Qi and/or Blood vacuity

c) **Level** – is represented as the location of the ECG elements in the vertical dimension on the greed in regards of the corridor of norm denoted by Π.

Clinical significance:
- Floating – exterior syndrome, interior and/or Yin vacuity
- Average - normal or requires clinical correlation
- Rootless – interior and/or Yin vacuity
- Sunken – interior syndrome, pathogen accumulation, Qi and/or Blood stagnation, Yang vacuity

d) **Rhythm** – Represents the regularity of the beats.

Clinical significance:
- Regular - normal or requires clinical correlation
- Regularly irregular – Wind stirring the interior, pathogenic factor accumulation, Qi and/or Blood stagnation, Qi and/or Blood vacuity
- Irregular - Wind stirring the interior, pathogenic factor accumulation, Qi and/or Blood stagnation, Qi and/or Blood vacuity
- Skipping - Wind stirring the interior, pathogenic factor accumulation, Qi and/or Blood stagnation, Qi and/or Blood vacuity

e) **Shape** – The presence, size, and the shape of ECG elements and their relationships to each other.

Clinical significance:
- String-like (Wiry) - Qi stagnation, Blood and/or Yin vacuity, phlegm accumulation
- Slippery - Dampness, phlegm and/or other pathogenic factor accumulation
- Thin (Thready) - Blood and/or Yin vacuity
- Rough (Choppy) - Qi, Blood and/or Yin vacuity, Blood and/or Qi stagnation

A pulse description combining all five pulse qualities, correlated with the tongue observation, and symptomatic clinical presentation offers a foundation to formulate a diagnosis and treatment plan. These findings result in an acupuncture protocol, herb-formula, or nutraceutical recommendation.

Chapter 8

Mobile ECG Based Pulse Case Studies

Introductory case study:
A post-menopausal female presents an occasional pounding headache in her left eye. Her symptoms are worse with low atmospheric pressure, may be accompanied by nausea, irritability, and tense neck worse on the left, difficulty to tolerate stuffy hot air. She reports occasional epigastric pain caused by stress and "rabbit pellets" constipation.
TCM Tongue: Pink, redder tip, Stomach crack, thin white coat.
TCM Pulse: Moderate, slightly uneven, rootless, regular, string-like (see example 1 pulse EKG and its interpretation on next page).

ICD 11 TCM Diagnosis: Migraine headache due to SF52 Liver yang ascendant hyperactivity pattern (TM1).
TCM Treatment plan: pacify the Liver to subdue Yang.

Discussion: patient symptoms (including tongue diagnosis) may not alone support an official diagnosis of Yin or Blood vacuity. However, by the definition, in the conditions of Liver Yang rising there always is an underlying Yin and Blood vacuity what the most clearly can be seeing in this rootless pulse. This example illustrates the value of ECG pulse diagnostic tool technique. It can identify early-stage functional disorders before disorders manifest structurally and become evident on the tongue appearance.

According to the WHO manual of standard terminology: "pacify the liver to subdue yang means: a therapeutic method to treat ascendant hyperactivity of liver yang by using yin blood nourishing medicinals and heavy mineral and shell medicinals."
On the curios note: This TCM treatment method correlates with the Western medical prescription of magnesium (heavy mineral) for this type of migraine.

1. Example: Moderate, slightly uneven, rootless, regular, string-like.

Western interpretations: providing no cardiac complaints are present:
Reduced T wave – may mean hypomagnesaemia and/or hypokalemia.

TCM Interpretations:
Rate - is 74 beats per minutes that is within range of normal 60-80 therefore can be described as moderate.
Force – the vertical strike of R wave is within range of normal but not on every beat, as well the base line waves randomly therefore it can be described as slightly uneven.
Level – S wave is not extending downward therefore this pulse can be described as rootless.
Rhythm – beats are positioned at even distance from each other therefore pulse is regular.
Shape – since the only R wave is prominent and palpable we can describe pulse as string-like.

Clinical significance:
Rate - is slightly faster than an ideal therefore we can assume the presence some degree of heat developing.
Force - normal Qi strength, unevenness may mean some wind.
Level - pulse is rootless which allows the Yang/heat to rise. In western medicine, a deficiency of kalium (potassium) is evident by a reduced T wave. Reduced or depressed ST interval points to intracellular dehydration. Dehydration correlates with the TCM concept of yin vacuity. These factors make the case that the pulse trends towards deficient heat.
Rhythm - beats are positioned at equal intervals, so the pulse is regular.
Shape - string-like in TCM may indicate Qi stagnation, tense personality with tense sinews. In Western medicine magnesium deficiency may manifest as a tissue tension, pain, headache and/or spasms. These symptoms mirror the TCM meaning of a string-like pulse.

1. **Example**: Moderate, slightly uneven, rootless, regular, string-like.

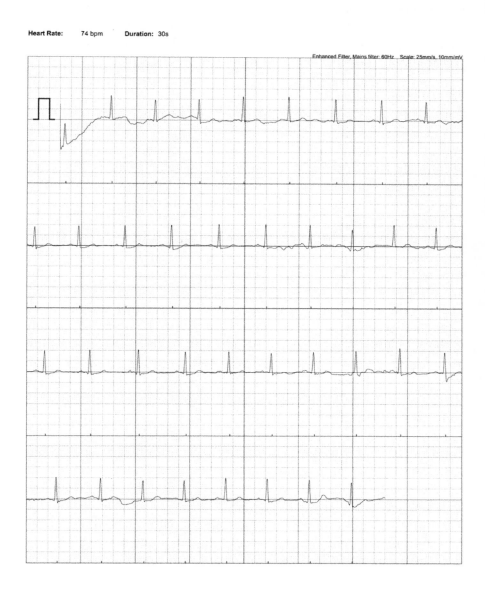

2. **Example:** Slow, vacuous on *sun*, rootless, regular, thin.

Western interpretations providing no cardiac complaints are present:
   The second row presents a downward premature ventricular contraction (PVC) spike. Occasional single PVC is fairly common in general population and does not raise cardiac concerns.

TCM Interpretations:
Rate - is 52 beats per minute that is below the range of normal 60-80. It is described as slow.
Force – the vertical strike of an R wave is reaching half of the range of normal. This dependent on the gender, age and physical condition of the patient may mean pulse vacuity.
Level – S wave is not extending downward therefore the pulse can be described as rootless.
Rhythm – regular, beats are mostly positioned at an even distance from each other with the exception of one spike of PVC, which requires more observation. This is not enough to claim rhythm disturbances.
Shape – there is a reduced or depressed P wave that is evident of a weakened performance of the atrial contraction. Upon palpation pulse is short and only mid-level can be felt. This in turn reduces amplitude of the QRS complex that in TCM is felt as a thin slightly vacuous pulse.

Clinical significance:
Rate - is slower than an ideal and may mean presence of cold.
Force - is half the normal signifying Qi deficiency, but some unevenness points us toward Qi stagnation as well.
Level - is not extending downward so pulse is not reaching deep level, which may mean interior deficiency.
Rhythm - presence of PVC may mean Wind pathogen lurking.
Shape - thin pulse maybe indicative of blood deficiency, which would explain formation of interior Wind that is causing PVC. As well this pulse has depressed P wave indicative of Lung weakness that along with rootlessness maybe indicative of difficulties for Kidneys to grasp Lung Qi leading to dyspnea and fatigue.

2. Example: Slow, vacuous on *sun*, rootless, irregular, thin.

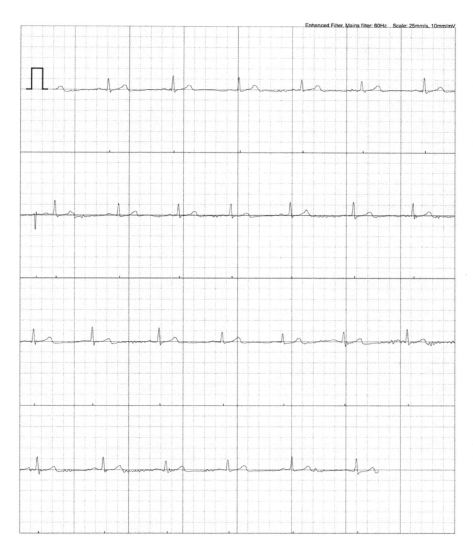

3. Example: Moderate, smooth, average, regular, vacuous on *chi*.

Western interpretations providing no cardiac complaints are present:
Reduced T wave – may mean subclinical hypokalemia and/or hypomagnesaemia.

TCM Interpretations:
Rate - is 72 beats per minute that is within the range of normal 60-80 therefore can be described as moderate.
Force - the vertical strike of R wave is 2/3 of range of normal and maybe just a physiologic variance dependent on gender of the patient or may require symptomatic correlation. Pulse has smooth flow.
Level - S wave is extending downward therefore this pulse has a good root.
Rhythm - beats are positioned at even distance from each other therefore pulse is regular.
Shape - there is no visible T wave and on palpation this pulse is short and absent on left Chi position.

Clinical significance:
Rate - is moderate so the temperature is normal.
Force - is within the normal so there is no Qi deficiency.
Level - pulse is nicely rooted, interior is strong.
Rhythm - regular, Qi moves smoothly.
Shape - there is no visible T wave. On palpation this pulse feels short, extremely vacuous or absent on *chi* position, and may signify Kidney deficiency or issues in the parts of the body governed by Kidney or Bladder meridians. Most likely due to some degree of hypomagnesaemia and/or hypokalemia as evident by greatly reduced T wave. Clinically this may manifest as back stiffness.

3.   Example: Moderate, smooth, sunken, regular, vacuous on *chi*

Heart Rate: 72 bpm   Duration: 30s

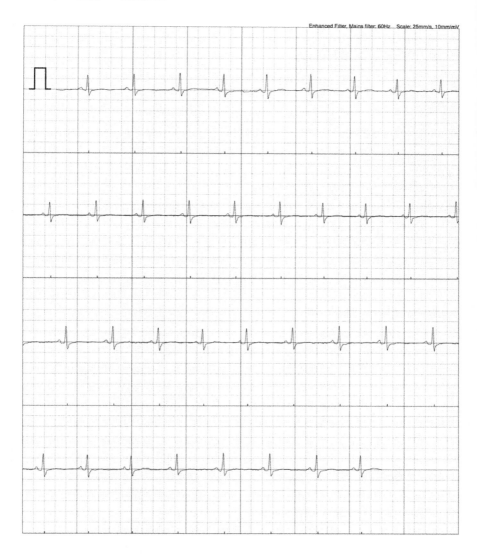

4. **Example:** Moderate, vacuous, uneven, rootless, regular, rough.

Western interpretations: this pulse most likely will manifest as palpitations and patient will have a western diagnosis, if not refer to Western MD.

TCM Interpretations:
Rate - is 73 beats per minutes that is within the range of normal 60-80, therefore can be described as moderate.
Force - the vertical strike of R wave is fluctuating to about 2/3rds of the height of the range of normal signifying average amount of Qi. But the base line is moving up and down so pulse maybe described as seriously uneven.
Level - Q & S waves are not extending downward therefore this pulse can be described as rootless.
Rhythm - systolic beats are positioned at even distance from each other therefore pulse is regular.
Shape - the only wave that is clearly distinguishable is R, remaining waves appear to be random in location, amount and shape. This pulse on palpation feels like rough (choppy).

Clinical significance:
Rate - is moderate though trending to warmer side.
Force - unstable height may mean Qi stagnation.
Level - is moving up and down that is evident by an unstable T wave. A depressed ST interval and a low T wave illustrates a possible potassium deficiency that means intracellular dehydration. The absence of a root presents some degree of yin deficiency. A low and unstable pulse line on the middle level means blood deficiency. This pulse points to interior Wind. It moves up and down, primarily on the *guan* position that may signify the Liver overacting on Spleen/Stomach pattern and manifest as alternating bowels.
Rhythm - pulse is regular.
Shape - on palpation this pulse feels rough (choppy) that is a classic definition of a blood deficient pulse. This provides the environment for the formation of internal wind, which may manifest as tremors.

4. **Example:** Moderate, vacuous, uneven, rootless, regular, rough.

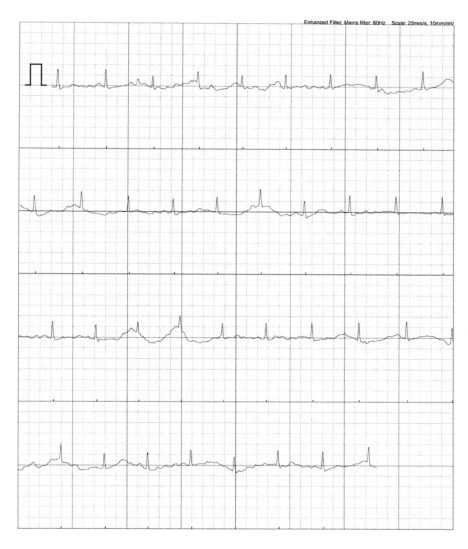

5. Example: Slow, vacuous, sunken, regular, slippery.

Western interpretations providing no cardiac complaints are present:
An increased T wave – may indicate a subclinical hyperkalemia and/or hypermagnesemia.

TCM Interpretations:
Rate - is 54 beats per minutes that is below the range of normal 60-80, therefore can be described as slow.
Force - the vertical height of R wave is $1/3^{rd}$ of the range of normal so pulse is vacuous.
Level - S wave is extending well downward and the base line prolapses below the bottom of Π line, therefore this pulse can be described as sunken/deep.
Rhythm - beats are positioned at even distance from each other, therefore pulse is regular.
Shape - the height of R and T waves is almost the same therefore this pulse can be described as slippery.

Clinical significance:
Rate - is slow that may be an indication of cold accumulation.
Force - is low that points to Qi deficiency.
Level - is deep, in western medicine, a hyperkalemia evident by an increase T wave interval that maybe illustrating swelling of the tissue. This correlates with the accumulation of dampness in TCM.
Rhythm - beats are positioned at equal intervals, so pulse is regular.
Shape - with 2 prominent elevations, this is a slippery pulse. In general, this pulse signifies Qi deficiency with an accumulation of damp cold.

5. Example: Slow, vacuous, sunken, regular, slippery.

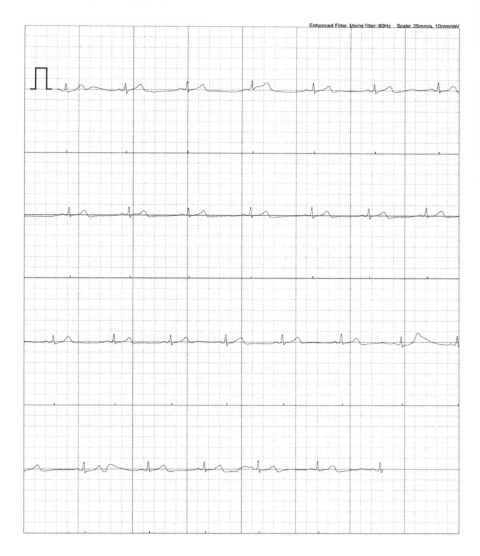

6. Example: Moderate, replete, floating, regular.

Western interpretations providing no cardiac complaints are present:
Slightly depressed ST interval may mean hyperventilation.

TCM Interpretations:
Rate - is 75 beats per minutes that is within range of normal 60-80 therefore can be described as moderate.
Force - the vertical strike of R wave is well above upper level of the range of normal therefore pulse is replete.
Level - S wave is not extending downward therefore this pulse can be described as rootless, in combination with replete in height it constitutes a floating pulse.
Rhythm - beats are positioned at even distance from each other therefore pulse is regular.
Shape – unremarkable.

Clinical significance:
Rate - is slightly faster than an ideal therefore we can assume that person may have some degree of heat developing.
Force - is above the range of norm that may signify presence of a pathogen.
Level - is floating indicating an exterior invasion, or maybe associated with heat accumulation evident by a high R wave. This could point to heat in the Liver overacting on the Spleen. This may cause dampness accumulation in the Spleen/Stomach *zang-fu* evident by depressed ST interval.
Rhythm – beats are positioned at equal intervals, so the pulse is regular.
Shape - is unremarkable.
In general rate is not fast enough to point to full heat. In combination with other factors this could indicate Phlegm accumulation in the Middle Burner.

6. Example: Moderate, replete, floating, regular.

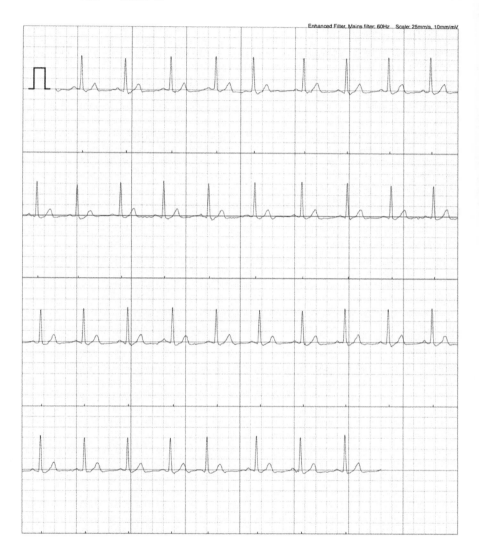

7. Example: Moderate, vacuous, sunken, irregular, thin.

Western interpretations there will be palpitations, may be POTS, possibly other cardiac complaints present.

TCM Interpretations:
Rate - is 66 beats per minutes that is within range of normal 60-80 therefore can be described as moderate.
Force - the vertical strike of R wave is barely visible therefore, this is vacuous pulse.
Level - waves below and above the base line are almost the same in size therefore this pulse can be described as sunken.
Rhythm - though R wave is positioned at regular intervals there are PVC's present, therefore pulse is irregular.
Shape - very low amplitude of heart contractions is able to eject small amount of blood that will produce thin pulse.

Clinical significance:
Rate - is slow, which may point to some cold accumulation or extreme Qi vacuity.
Force - is extremely low illustrating severe Qi vacuity.
Level - is sunken concluding an interior condition.
Rhythm - irregular PVC's reveal Qi stagnation and Wind in the interior.
Shape - is thin the may be a sign of Blood vacuity.

7. **Example:** Moderate, vacuous, sunken, irregular, thin.

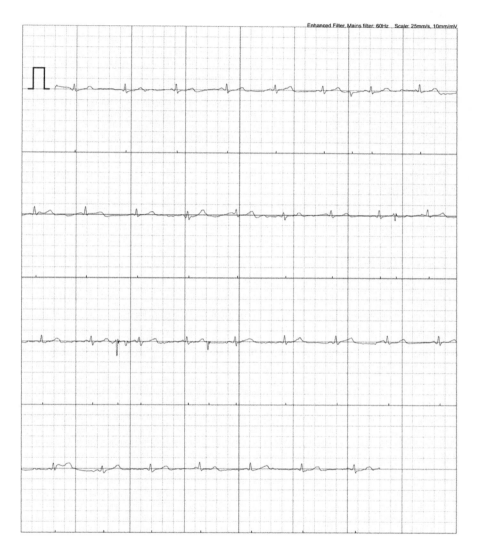

8. Example: Fast, average, slightly uneven, vacuous on *sun*

Western interpretations providing no cardiac complaints are present:
Reduced P wave may mean ischemia that may manifest as shortness of breath and low stamina.

TCM Interpretations:
Rate - is 93 beats per minutes, which is above the range of the normal of 60-80, therefore the rate is described as fast.
Force - the vertical strike of the R wave is within range of normal but not on every beat, therefore it can be described as slightly uneven.
Level - S wave is extending downward therefore this pulse can be described as average.
Rhythm - beats are positioned at even distance from each other therefore pulse is regular.
Shape - P wave is depressed that may be described as vacuous on *sun* position pulse; other vice unremarkable.

Clinical significance:
Rate - is fast which indicates heat.
Force – R wave is within the normal showing average amount of Liver Qi. There is some unevenness, which points us toward Qi stagnation.
Level - is present on all levels therefore is average and may need clinical correlation.
Rhythm - beats are positioned at equal intervals, so pulse is regular.
Shape - the most prominent feature is depressed P wave. This may mean Qi deficiency in the Upper Burner.
In general fast rate and P wave vacuity may signify deficient heat in the lungs and/or the heart.

8. Example: Fast, average, slightly uneven, vacuous on *sun*

Heart Rate: 93 bpm   Duration: 30s

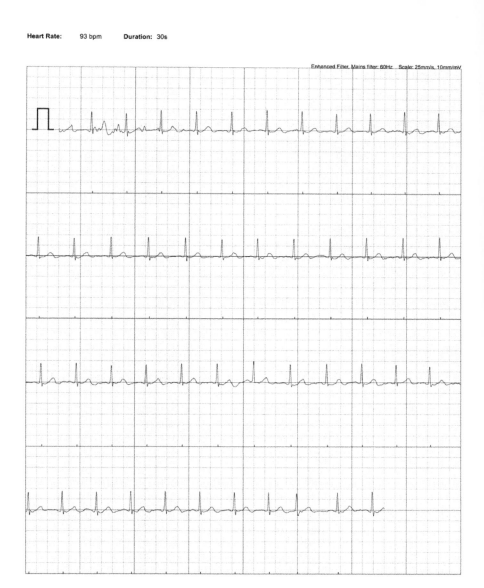

53

9. Example: Moderate, vacuous, sunken, regular, string-like, thin

Western interpretations providing no cardiac complaints are present:
Reduced T wave may mean subclinical hypokalemia and/or hypomagnesaemia. Reduced P wave may mean ischemia that may manifest as shortness of breath.

TCM Interpretations:
Rate - 70 beats per minutes is within range of normal 60-80 therefore can be described as moderate.
Force - the vertical strike of R wave is at half height of range of normal therefore it can be defined as slightly vacuous.
Level - S wave is not extending downward therefore this pulse can be described as rootless.
Rhythm - beats are positioned at even distance from each other, therefore pulse is regular.
Shape - since only R wave is visible and therefore palpable we can describe pulse as string-like. Taking only half of blood level and being rootles it is thin (thready).

Clinical significance:
Rate - contrary to example 1 this rate has no heat in it.
Force - is indicative of Qi deficiency. Some unevenness points toward Qi stagnation.
Level - is not extending downward so pulse is not reaching deep. This correlates with an understanding of yin deficiency and/or interior deficiency in TCM.
Rhythm - pulse is regular.
Shape - has only one elevation, which feels as a string-like beat on palpation. This points to Liver Qi stagnation. The pulse is thin that points to blood deficiency. A reduced P wave may mean Qi deficiency in the Upper Burner. A reduced T wave may mean Qi deficiency in the Lower Burner.
In general this pulse could be representative pulse of Xiao Yao San formula.

9. Example: Moderate, vacuous, rootless, regular, string-like, thin.

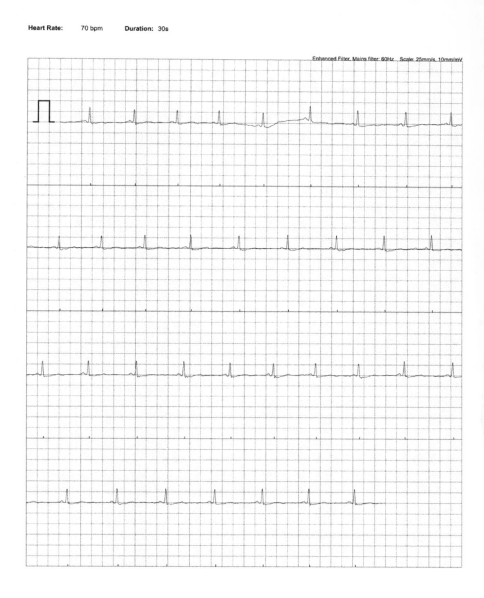

10. Example: Slow, vacuous, rootless, regularly irregular, slippery.

Western interpretations: there is clear cardiac complaint,
Reduced misshaped R wave means that bundle of His branch block is present. This also may indicate subclinical hyperkalemia that is supported by taller T wave.

TCM Interpretations:
Rate - is 52 beats per minutes is below range of normal 60-80
Force - the vertical strike of an R wave is below half the height of the range of normal therefore it is Qi vacuous pulse
Level - ST interval is inverted that can be described as sunken pulse.
Rhythm - R wave beat is mostly regular but has irregular double shape so it can be described as regularly irregular. Single occasional PVC's is insufficient to draw any conclusions and requires more observation.
Shape - with all waves being almost the same height it will feel as slippery on palpation. Reduced P wave illustrates vacuous pulse on *sun* position signifying Upper Burner deficiency.

Clinical significance:
Rate - is slow indicating cold accumulation and/or Yang deficiency.
Force- is low pointing to Qi deficiency.
Level - is low meaning an interior condition.
Rhythm - irregularity means Qi stagnation.
Shape - slippery correlates with dampness accumulation,
In general this pulse may illustrate deep-rooted cold phlegm accumulation in the Middle Burner and vacuity in the Upper Burner. Upper Burner vacuity may manifest as insomnia, and palpitations. Cold-phlegm in the Middle Burner, in the advanced stage may manifest as lymphoma or other type of tumors.

10. Example: Slow, vacuous, rootless, regularly irregular, slippery.

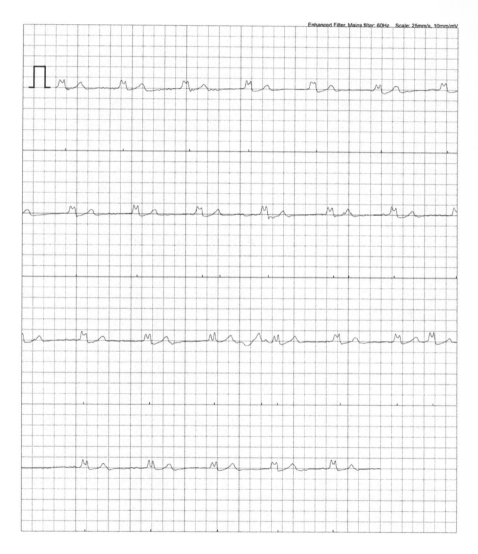

11. Example: Racing, vacuous, rootless, irregular, thin.
Western interpretations: IT IS A RED FLAG – Atrial fibrillation.

TCM Interpretations:
Rate - is 116 beats per minute (well above the range of normal 60-80). This can be described as racing pulse.
Force - the vertical strike of R wave is less than a half of range of normal therefore pulse is vacuous.
Level - S wave is not extending downward therefore this pulse can be described as rootless.
Rhythm - beats are irregular.
Shape - irregular shape of waves can be described as rough.

Clinical significance:
Rate - is racing that means extreme heat or fire.
Force - is below normal representing a significant Qi deficiency.
Level - is not extending downward, therefore the pulse is not reaching a deep level. This points to Yin deficiency.
Rhythm - is irregular meaning Qi stagnation. Qi stagnation creates an accumulation of heat.
Shape - is thin indicating blood deficiency.
In general, this pulse illustrates deficient heat steering internal wind, caused by heat in the blood due to blood and yin deficiency.

11. Example: Racing, vacuous, rootless, irregular, thin.

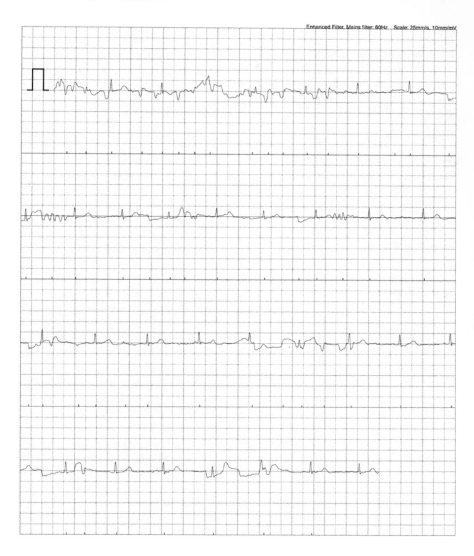

12. Example:

Clinically, I encountered 2 cases in which I could not record a mobile ECG. Neither of the patients had life-threatening health concerns.

In one case, the patient's pulse was fast, vacuous, rootless, regular, and minute. The patient was too dry for a recordable strength of the electric signal to be transmitted. The patient was extremely, Qi and Yin deficient, with severe Qi stagnation.

In the other case, upon palpation, the patient's pulse was slow, vacuous, sunken, irregular, and slippery. The patient's cardiology had long QT syndrome. The clinical TCM presentation was blood stagnation and damp accumulation in the lower burner.

NOTE: A powerful Internet reception signal is essential. If it is below 4 G, the recording of Qi deficient pulses becomes difficult.

Conclusion

There are many pulse variations. This book is an attempt to correlate palpatory findings with the author's understanding of TCM theory to interpret the objective mobile ECG visual diagram to describe a TCM pulse. Readers are encouraged to apply their own knowledge to correlate symptoms, pulse palpation, and tongue observation with an ECG recording before drawing clinical conclusions.

The author invites commentary and scientific discussion on the subjects presented in this book. Email: GR@AsianTherapies.org

Sincerely,
Galina V. Roofener

Chapter 9

References

1. Becker DE. Fundamentals of electrocardiography interpretation. *Anesth Prog.* 2006;53(2):53-63; quiz 64.

2. Yasin OZ, Attia Z, Dillon JJ, et al. Noninvasive blood potassium measurement using signal-processed, single-lead ecg acquired from a handheld smartphone. *J Electrocardiol.* 2017;50(5):620-625.

3. Ching-HsingLuo, Chen-YingChung. Non-invasive holistic health measurements using pulse diagnosis: II. Exploring TCM clinical holistic diagnosis using an ingestion test. *European Journal of Integrative Medicine.* Volume 8, Issue 6, December 2016, Pages 926-931

4. Karen Bilton, Leon Hammer, Chris Zaslawski. Contemporary Chinese Pulse Diagnosis: A Modern Interpretation of an Ancient and Traditional Method *J Acupunct Meridian Stud* 2013;6(5):227e233

5. Yun-Kyoung Yim et al. Gender and Measuring-position Differences in the Radial Pulse of Healthy Individuals. *J Acupunct Meridian Stud* 2014;7(6):324e330

6. Yuh-Ying Lin Wang et al. Past, Present, and Future of the Pulse Examination *Journal of Traditional and Complementary Medicine* Vol. 2, No. 3, pp. 164-185

7. Nathalia Gomes Ribeiro Moura, Arthur Sa Ferreira. Pulse Waveform Analysis of Chinese Pulse Images and Its Association with Disability in Hypertension. *J Acupunct Meridian Stud* 2016;9(2):93e98

8. Jibing Gong et al. Low-cost and Wearable Healthcare Monitoring System for Pulse Analysis in Traditional Chinese Medicine. *IEEE 7th International Conference on Mobile Adhoc and Sensor Systems,* MASS 2010, 8-12 November 2010, San Francisco, CA, USA

9. Yu-Wen Chu et al. Using an array sensor to determine differences in pulse diagnosis—Three positions and nine indicators. *European Journal of Integrative Medicine* 6 (2014) 516–523

10. Cornelis S. van der Hooft et al. Drug-induced atrial fibrillation Journal of the American College of Cardiology Volume 44, Issue 11, 7 December 2004, Pages 2117-2124

11. https://www.merckmanuals.com/professional/cardiovascular-disorders/arrhythmias-and-conduction-disorders/long-qt-syndrome-and-torsades-de-pointes-ventricular-tachycardia

12. https://crediblemeds.org/

13. https://www.uspharmacist.com/article/drug-induced-qt-prolongation

14. https://litfl.com/ecg-library/

15. https://en.wikipedia.org/wiki/Organophosphate

16. Jessica Shank Coviello, LWW. ECG Interpretation: An Incredibly Easy Pocket Guide. Wolters Kluwer Health.

17. Pyers, Clare. 2016 Integrative TCM Guide - Pathology: Interpreting Blood Tests into a Chinese Medicine Framework. Discover Chinese Medicine.

18. Keown Daniel. 2018 The Uncharted Body: A New Textbook of Medicine

 Galina V. Roofener is a NCCAOM board certified and licensed in Ohio and Florida to practice Oriental medicine.

Currently, Galina works for the Cleveland Clinic as an acupuncturist and Chinese herbalist. Cleveland Clinic was one of the first Western medical hospitals to open a hospital-based full-spectrum custom compounded Chinese herbal clinic in the U.S.

She serves on the State Medical Board of Ohio's Acupuncture and Oriental Medicine Advisory committee and is a member of the Hospital-Based Practice Task Force Committee advisory panel for the National Certification Commission on Acupuncture and Oriental Medicine.

Galina has over 30 years of experience in health care, and is a published author, avid teacher, and renowned speaker. She is an NCCAOM approved PDA provider. More information can be found at www.AsianTherapies.org

CPSIA information can be obtained
at www.ICGtesting.com
Printed in the USA
LVHW111144180819
628041LV00001B/193/P